A TALE NEVER TOLD:
The revealed brief history of earth nobody knows.

Prof Ryan A. Smith

All rights reserved. No part of this publication may be reproduced, distributed, or transmitted in any form or by any means, including photocopying, recording, or other electronic or mechanical methods, without the prior written permission of the publisher, except in the case of brief quotations embodied in critical reviews and certain other noncommercial uses permitted by copyright law.

Copyright © Prof Ryan A. Smith, 2022.

Table of contents

Introduction

Chapter 1
Origin of this earth

Chapter 2
Birth of the sun and moon

Chapter 3
Formation of the blue seas and oceans

Chapter 4
Beyond our universe.

Chapter 5
9 truths about extraterrestrials

Introduction

How did the universe come to be? Since the beginning of time, people have asked this question and have received answers from a wide range of sources, including science, religion, tradition, philosophy, and mysticism. Although this issue does not appear to be amenable to scientific measurement, it has nonetheless inspired scientists to make some fascinating discoveries, including the Big Bang, the idea of inflation, and the discovery that the majority of the universe is composed of invisible dark matter and dark energy, and more. Of course, scientists cannot assert that they are aware of the whole truth. But we can approach the issue scientifically and see what we discover. How do we go about doing that? We start by looking at the data. We have a lot more information now than people in earlier eras who asked the same question did because of modern technology. The data can then be analyzed and arranged coherently, and an attempt is made to derive an answer using scientific methods and techniques. This process and its main findings will be described in this book.

The concept of creation takes on a particular meaning in a scientific context, not to be confused with the concept of "creation out of nothing" that we find in metaphysics or monotheist theologies. In its narrow and most commonly used sense, it means a specification of the state of the universe at some initial time, together with the laws of physics that have evolved this initial state up until today. The initial state may or may not be approximately classical or quantum and the laws of evolution may involve quantum mechanical equations or classical equations. Sometimes the specification of the initial state is only statistical, chosen from some ensemble of states with a prescribed probability. In this case, the idea of one initial state is replaced by the set of possible initial states and the probability distribution on it. Even in Stephen Hawking's description of the universe expanding from "nothing," the quantum wavefunction's beginning conditions must be specified. So we need to think about what could have been the first circumstances before we can talk about creation. As a result, the scientific definition of "creation" is essentially a mathematical explanation of the starting circumstances of a "natural beginning" or an "emergence from something" in terms of equations.

Chapter 1

Origin of this earth

Human thinking and scientific inquiry have long been engaged with issues related to the origins and character of Earth. Understanding the history and dynamics of the planet might help us better manage Earth's resources, forecast changes in climate, and foresee disasters like earthquakes and volcanoes. The key problems that make up the frontier of Earth science at the beginning of the 21st century are summarized in this study via a series of questions.

Earth is a dynamic planet. Mountain ranges and the seafloor are continuously being built and destroyed, earthquakes tear along plate borders, and volcanoes erupt streams of molten lava. For a very long time, earth scientists have been interested in understanding the past and foretelling the future of this dynamic planet. Earth scientists have made significant advancements in their knowledge of Earth's functioning during the last forty years. Scientists can now better grasp how internal processes on Earth influence the planet's surface, how life may persist across billions of years, and how interactions between geological, biological, atmospheric, and oceanic processes result in climate—and climatic change—thanks to ever-improving technologies. The National Research Council formed a committee to make and investigate proposals at the request of the U.S. Department of Energy, National Aeronautics and Space Administration, National Science Foundation, and U.S. Geological Survey regarding big issues in Earth science. The 10 "large picture" Earth scientific challenges being explored today are described in this report, which is the outcome of the committee's discussions and information gathered from the Earth science community. The answers to these basic questions might significantly advance our knowledge of the world we live in and our environmental management techniques.

1. How did the Earth and the other planets originate?

The planets and moons that make up the Solar System range from the rocky inner planets to the gas giants Jupiter, Saturn, Uranus, and Neptune. The creation of models for the formation of the Solar System has been made possible by centuries of research into Earth, its surrounding planets, and meteorites. These models have been expanded by astronomical observations made with progressively stronger telescopes, spacecraft studies of asteroids, comets, and other planets, as well as geochemical investigations of meteorites and stardust. Although it is commonly accepted that the Sun and planets all formed from the same nebular cloud, nothing is known about how Earth's specific chemical makeup developed or why the other planets ended up being so unlike Earth and one another. Why, for instance, does Earth still possess the special qualities that make it possible for life to exist, such as the availability of water, unlike any other planet? Understanding of the formation of Earth and the Solar System will be furthered by new observations of Solar System entities and extrasolar planets and objects.

2. What occurred during the first 500 million years of Earth's history, or the "dark age"?

The current theory is that a planet the size of Mars collided with Earth during its creation, causing a massive cloud of debris to develop into Earth's Moon and a tremendous amount of heat to be released, causing the planet to melt. However, little is known about how the molten rock that resulted during the planet's early development transformed into the Earth that we know today. Understanding how Earth's atmosphere, seas, and distinct layers of the core, mantle, and outer crust originated requires knowledge of the first 500 million years of the planet's history or the Hadean Eon. Almost no one knows how quickly the surface environment changed, how the shift occurred, or when the environment was favorable enough to host life. Zircons, one of the oldest minerals on Earth, as well as the Moon and other planets are providing some hints that are progressively helping to paint a better image of the Hadean Eon. More innovations are sure to come in the future. Even the smallest pieces of ancient rocks and minerals may contain a wealth of information, and with focused effort, it is anticipated that many more examples of these materials will be discovered.

3. How does the interior of the Earth function, and how does this impact the surface?

Internal and surface processes on aging and cooling planets progressively alter. The characteristics of Earth's surface and atmosphere are greatly influenced by the manifestations of changes inside the planet's interior, such as the growth of mountains and volcanoes. Most of the rock in the Earth's mantle, which is the thick layer between the core and crust and is subject to very high pressure and temperatures, is known to behave much like a viscous liquid. Direct research, however, is almost impossible in this huge interior. Seismic wave, geomagnetic, and gravity data were taken at the surface have advanced knowledge of the Earth's underlying structure for more than a century. Even yet, researchers are just now starting to analyze the differences between Earth and other planets, as well as the linkages between the

Earth's core, magnetic field, mantle, and surface.

4. Why do continents and plate tectonics exist on Earth?

Understanding the nature of the continents—the physical qualities that make Earth livable for land-dwelling life—has been a significant emphasis of Earth research. The formation of the continents: how and when? How do they differ now? Why do South America's and Africa's Atlantic coasts resemble jigsaw pieces? Since it became the dominant paradigm for geology about 40 years ago, plate tectonics—the theory that describes the Earth's outermost layers in terms of a few huge rigid plates moving relative to one another—has provided several ground-breaking revelations. It is now understood that the movement of tectonic plates and interactions at their borders are the primary causes of earthquakes, volcanic eruptions, the development of mountains, and the gradual movement of continents across the surface of the Earth. Although the idea of plate tectonics explains many of Earth's surface characteristics, it remains unknown why the planet has plates or how plate tectonics is related to the planet's copious water supply, continents, and the presence of life. Such problems will need better models, research on contemporary plate boundaries, and comparisons with other planets to be answered.

5. How do the characteristics of materials affect Earth's processes?

The little differences are what count. The composition of the materials that make up the globe, even down to the tiniest specifics of their atomic structures, is now understood to be the driving force behind large-scale phenomena on Earth, such as plate tectonics.

Studying Earth's materials is difficult due to the extreme pressures and temperatures of the planet's interior, the massive size of the globe and its structures, the length of geological time, and the wide variety of materials. New analytical tools and sophisticated computing capabilities are advancing the study and simulation of Earth materials at the atomic level, and they promise to improve predictions of how these material properties will affect planetary processes. However, breakthroughs in this field are now within reach.

6. What influences climate change and by how much?

It is commonly acknowledged that emissions of CO_2 and other greenhouse gases are at least partially to blame for the increase in Earth's mean global surface temperature since the start of the industrial period. Given the potentially devastating effects of global warming, it is imperative to ascertain how much of the warming is a result of human activity and what can be done to stop it. Both concerns have a significant bearing on earth science. The geological record has shown that the planet's climate has had an odd history of unpredictability and stability. For the last 10,000 years, the climate has been reasonably steady and beneficial for life, and it has been this way for more than 3 billion years. However, geological data also demonstrates that significant climatic shifts may take place over timescales as brief as decades or centuries. How, given how quickly it may change, can Earth's climate stay mostly steady throughout time? New knowledge about Earth's climate is being gained through studying times when the globe was exceptionally cold, heated, or changed particularly swiftly. In the future, observations of old rocks may help to more accurately anticipate the scope and effects of climate change.

7. How has the world changed as a result of life, and vice versa?

Scientists are aware that the existence of life has had an impact on the makeup of the Earth's atmosphere, particularly its high oxygen content. Life is an unseen but potent chemical force at the microscopic level. Organisms catalyze processes that would not occur without them and speed up or slow down other reactions. These responses may result in changes that have an impact on the whole planet when they are multiplied over long periods by massive biomass. The geologic development of Earth as well as disastrous occurrences like meteorite strikes have undoubtedly influenced the emergence of life. The reasons for extinctions and significant evolutionary shifts, however, are yet unknown. How many of their causes were geological as opposed to biological? There is an ongoing discussion on how exactly geological events have impacted evolution and how much influence life has had on climate. A significant problem is figuring out how life and the processes that form the land interact.

8. Is it possible to foresee earthquakes, volcanic eruptions, and the damage they cause?

The abrupt and dangerous effects of the Earth's interior's slow motions are most often seen in earthquakes and volcanic eruptions. As human populations cluster more and more in earthquake- and volcano-prone regions, the need to foresee such disasters has increased. Geologists are advancing toward volcanic eruption prediction skills, in great part due to sensitive new instruments and increased knowledge of causes. Although there has been improvement in assessing the likelihood of future disasters, it may never be possible to foretell the precise time and location where an earthquake will hit. Determining the beginning and ending points of fault ruptures, modeling the amount of shaking that may be anticipated close to significant earthquakes, and lengthening the time that warnings are given after a catastrophic earthquake starts are all ongoing difficulties.

9. How do fluid movement and flow impact the world around people? Many of the most precious resources on Earth are present, and their location is dictated by fluid movement and transport mechanisms. Understanding fluids, both above and below ground, is essential for evaluating and extracting minerals, oil, gas, and groundwater as well as for securely dumping the garbage. Some of the fluid-related scientific topics also have implications for predicting earthquakes, the climate, the development of continents, the behavior of volcanoes, and the characteristics of Earth's materials. A whole new perspective on the processes affecting Earth's fluids is now possible because of new experimental methods, equipment, and data taken from

the air and space. Kilauea volcano, Hawaii, 9/19/84, nevertheless USGS, C. Heliker as the source. Many remain unanswered. The ultimate goal, to create mathematical models that can forecast how natural systems will behave in the future, is still a ways off, but achieving it will be essential to making wise judgments about the future of the resources and land that sustain us.

Chapter 2

Birth of the sun and moon

The sun

Gravity brought dust and gas together in a vast region of space to form the early solar system. Out of the huge mass, the sun emerged first, followed closely by the planets. But how did the brightest star in our sky emerge from a sea of whirling particles?

The finest physics laboratory in our solar system is the sun, which is both scary and beautiful, according to Sabrina Savage, project scientist for NASA's Hinode at the Marshall Space Flight Center in Huntsville, Alabama.

Despite seeming to be empty, space is filled with gas and particles. The majority of the substance was hydrogen and helium, but part of it also included star-star violent deaths' residual debris. The solar system originally began to develop some 4.5 billion years ago as waves of energy moving through space compressed clouds of these particles closer together. Gravity then drove them to collapse in on themselves and begin to spin. The cloud became flattened into a pancake-like disk as a result of the rotation. The material gathered in the center to create a protostar, which would later develop into the sun.

Astronomer Jared Holmes described a comparable early sun as having a rotating disk surrounding it, describing it as an "essential ingredient" in creating planets because "it allows the material hang around long enough for the planet formation process."

A ball of hydrogen and helium that wasn't yet driven by fusion served as the early protostar. Over tens of millions of years, the material's temperature and pressure rose, triggering the hydrogen fusion that powers the sun today.

According to NASA, it takes a star the size of our sun 50 million years to grow from the start of the collapse to maturity. "Our sun will continue to develop at this stage for around 10 billion years."

Not all of the cloud from which the sun was created was consumed in its creation. Planets were created from the remaining material, and what was left continued to revolve around the star. The sun is a star of normal size, neither too large nor too little. Given that it is neither enormous nor tiny or faint, its size makes it a great start to circle.

The sun will run out of hydrogen in a few billion years, causing it to grow out into a red giant with a radius that reaches Earth's orbit. It will also lose its core's helium via consumption. The sun will eventually fade away and transform into a white dwarf because the star will never be hot enough to burn the oxygen and carbon that are still there.
Of course, there were no human scientists to research the sun when it was created billions of years ago. Astronomers examine the many stars in the Milky Way to get knowledge about the life of the sun. These findings, when combined with models, may

help us learn more about the young age of our nearest star.

How hot is the sun today?

Different strata have different temperatures. So how hot is the sun exactly?

Without our enormous hot, incandescent ball of gas, life on Earth would not exist. But how hot is the sun exactly? That depends, I guess.

According to NASA, the sun's temperature ranges from its core, which is around 27 million degrees Fahrenheit (15 million degrees Celsius), to its surface, which is only about 10,000 degrees F (5,500 degrees C).

According to NASA Space Place, the sun emits more energy every 1.5 millionths of a second than all people use in a year.

(new tab opens) Here, we examine the temperatures of the sun's many layers and why they differ so much.

Gas and plasma make up the sun. Hydrogen makes about 92% of the gas. The sun would only be a huge hydrogen ball, similar to Jupiter if it were smaller. According to NASA Space Place, the sun's core's hydrogen is kept together by a strong gravitational field, causing tremendous pressure. Nuclear fusion is a process that occurs when hydrogen atoms collide with enough force to produce helium as a result of extreme pressure.

The sun's core reaches temperatures of roughly 27 million degrees Fahrenheit due to continuous nuclear fusion, which builds up energy. The sun's surface, atmosphere, and farther reaches receive the energy as it shines.

According to the educational website Study.com, the radiative zone is the region beyond the sun's core where temperatures vary from 12 million Fahrenheit (7 million Celsius) closest to the core to around 4 million Fahrenheit (2 million Celsius) in the outer radiative zone. According to Phys.org, a website that covers scientific news, there is no heat convection in this stratum. Thermal radiation, in which hydrogen and helium produce photons that travel a short distance before being reabsorbed by other ions, is used to convey heat instead. Before reaching the surface of the sun, light particles (photons) may travel through this layer for tens of thousands of years.

According to Study.com, the sun's convective zone extends for 120,000 miles (200,000 kilometers) beyond the radiative zone. The convection zone has temperatures of around 4 million degrees Fahrenheit. Like boiling water, the plasma in this layer travels in a convective manner, and hot plasma bubbles carry heat to the sun's surface.

Temperatures in the chromosphere, which is located above the photosphere, vary from around 11,000 degrees Fahrenheit (6,000 degrees Celsius) near the photosphere to about 7,200 degrees Fahrenheit (4,000 degrees C) a few hundred miles above.

Up to 500 times hotter than the photosphere, the sun's corona may reach temperatures of between 1.8 million to 3.6 million degrees Fahrenheit (1 to 2 million degrees Celsius). How, therefore, is the higher atmosphere of the sun hotter than the surface? It's a wonderful question, and scientists are perplexed by it. Thoughts concerning the source of the energy that warms the corona exist, but a conclusive answer has not yet been reached. Check out this article on "Why is the sun's atmosphere hotter than its surface?" to learn more about this solar riddle.

The Parker Solar Probe, which was launched in August 2018 and is presently in orbit around our star, will use one of its main objectives to look into why the corona contradicts stellar dynamic theories by being hotter than the photosphere.

The aircraft will travel through the sun's atmosphere while enduring extreme heat, often passing within 3.8 million miles of the sun's surface. While doing this, it will take pictures of the star and measure the corona to get crucial information on solar winds.

In 2021, the probe passed the sun at 364,621 mph, making it the fastest human-made object ever. According to NASA's Parker Solar Probe website, the Parker Solar Probe travels at a speed of 430,000 mph (700,000 kph) when it is closest to the sun.

It should come as no surprise that stars may be of diverse shapes, sizes, and hues and that they can even fluctuate in temperature. By looking at a star's color or spectral type, astronomers may learn a lot about its temperature.

The letters O, B, A, F, G, K, and M stand for the seven different spectral categories. The hottest stars are O and B stars, which mostly emit blue light with a significant amount of ultraviolet energy. The coldest class of stars, M-type stars are more conspicuous in red wavelengths while simultaneously producing a significant amount of infrared radiation.

According to the University of Central Florida, red stars are substantially colder than blue stars, with surface temperatures of about 3,000 K (4,940 degrees F/ 2,726 degrees C), whereas blue stars are believed to have surface temperatures of 25,000 kelvin (K) (44,540 degrees F/ 24,726 degrees C). Between them are yellow stars, like the sun, at 6,000 K (10,340 degrees F/

5,726 degrees C), white stars with temperatures around 10,000 K (17,540 degrees F/ 9,726 degrees C), orange stars with temperatures around 4,000 K (6,740 degrees F/ 3,726 degrees C), and white stars with temperatures around 10,000 K (17,540 degrees F/ 9,726 degrees C).

The size of the sun

The sun is the most enormous object in our solar system, but how big is it?

Although the sun is the greatest object in the solar system, how enormous is it concerning the other stars in the Milky Way galaxy? Among these tens of billions of stars, the sun is considered to be medium-sized.

The sun is almost an ideal sphere. Its polar and equatorial diameters are barely 6.2 miles apart. The sun's average radius is 432,450 miles (696,000 kilometers), giving it a diameter of around 864,938 miles.

According to NASA, you could align 109 Earths across the face of the sun. There are about 2,715,396 miles around the sun.

The sun may be the largest object in this vicinity, yet it pales in comparison to other stars. Red monster Betelgeuse is around 700 times larger and 14,000 times brighter than the sun.

"We have discovered stars with a diameter 100 times greater than the suns. Those stars are quite huge "According to NASA's SpacePlace website We've also seen stars that are just one-tenth as big as our sun.

C. Alex Young(opens in new tab), a solar scientist at NASA, estimates that if the sun were hollow, it would need around one million piles of earth to fill it.

The sun could be considerably bigger than originally believed. To pinpoint exactly where the moon's shadow would fall during a solar eclipse, engineer and solar eclipse researcher Xavier Jubier builds intricate models of solar and lunar eclipses. However, he discovered that accurate eclipse forms only made sense if he scaled up the sun's radius by a few hundred kilometers when he compared real images and historical observations with the models.

The radius of the star is not refined as accurately as needed by any missions, not even NASA's Solar Dynamics Observatory (SDO) or observations of the inner planets across the face of the sun.

Ernie Wright, a NASA researcher, told Space.com that "it's tougher than you think merely to put a ruler on these photographs and figure out how large the sun is - [SDO] doesn't have enough accuracy to nail this down." Similarly, it turns out that a measurement based on the Mercury and

Venus transits is not nearly as accurate as you'd want it to be.

According to Wright, data from several articles employing various techniques have varied by as much as 930 miles. If you want to see the upcoming solar eclipse, that may be an issue.

Yes, it won't make much of a difference for the majority of individuals, Jubier added. However, the danger increases as you approach the [eclipse] path's edge.

The moon.

According to Sara, "There used to be a lot of hypotheses about how the Moon was created, and one of the Apollo program's goals was to discover out how we came to have our Moon."

There were three hypotheses on how the Moon evolved before the Apollo missions, according to studies.

According to the capture hypothesis, the Moon was once a roving body (similar to an asteroid) that originated somewhere in the solar system and was drawn in by the gravity of Earth as it went by. On the other hand, the accretion hypothesis postulated that the Moon formed at the same time as the Earth. The fission theory concluded that Earth had been spinning so quickly that some material had finally broken free and started to circle the planet.

The giant-impact hypothesis is now the one that is most frequently accepted. According to this theory, the Moon was created after the Earth collided with a smaller planet that was around Mars' size. The Moon was created when the impact's leftover material gathered in an orbit around Earth.

Apollo mission data

Over a third of a tonne of rock and dirt from the Moon was carried back during the Apollo missions.

Sara explains that when the Apollo rocks were analyzed, they revealed striking chemical and isotopic similarities between the Earth and the Moon that point to a shared past.

We would anticipate that the Moon's composition would be significantly different from that of the Earth if it had been produced somewhere else and then drawn in by the gravity of the Earth.

"We would anticipate the kind and proportion of minerals on the Moon to be the same as on Earth if the Moon was produced at the same period as the Earth or if it broke off from it." However, there is a little difference.

Compared to comparable rocks on Earth, the minerals on the Moon have lower water content. The substance that develops fast at high temperatures is abundant on the Moon.

The big impact model was nearly universally accepted as a result of the extensive discussion that took place in the 1970s and 1980s.

Another significant source of information for researching the Moon's origins is lunar meteorites.

Sara continues, "In some respects, meteorites may teach us more about the Moon than Apollo samples since meteorites originate from all over the Moon's surface, but Apollo samples only come from one location near the equator on the near side the Moon.

Earth's biggest offshoot

Proto-Earth and Theia existed before Earth and the Moon.

The giant-impact hypothesis postulates a collision between these two bodies sometime in Earth's very early history.

Nearly all of Earth and Theia melted and reformed as one body during this violent collision, with a small portion of the new mass spinning off to become the Moon as we know it.

To get the closest match, scientists have experimented with impact modeling, changing Theia's size to test what happens at various sizes and impact angles.

Sara explains that due to their proximity to where the solar system was forming, "people are now gravitating towards the idea that early Earth and Theia were made of almost the same materials, to begin with."

This would also explain how similar the two bodies' compositions are: "If the two bodies had come from the same place and were made of similar stuff, to begin with."

the lunar surface

Because Earth and the Moon share similar mineralogy, it is possible to see Moon-like landscapes without traveling into space.

Sara explains, "If you look at the lunar surface, it appears to be a light grey with dark splotches." The rock known as anorthosite is a light gray color. It forms as molten rock cools and lighter materials rise to the top; the dark areas are made of basalt, another type of rock.

a moderating force

Our solar system is the only one in which a moon of such size exists.

The moon of Earth is nearly the same size as Mars, whereas the moons of other planets are much smaller, claims Sara.

If you take a look at other planets that are similar to ours, their orbits wobble quite a bit (the North Pole moves), and as a result, the climate is much more unpredictable.

The Moon has lessened polar motion and helped to stabilize Earth's orbit. This has contributed to the relatively stable climate on our planet.
How significant the Moon has been in enabling life to exist on Earth is a matter of considerable scientific debate.

Is there more than one moon on Earth?

There might be several things in Earth's orbit. To the best of our understanding, however, they are bodies that the planet has brought into its orbit; they are most likely captured asteroids. These natural satellites are likely only present in Earth's orbit for a brief period and do not have the same significant past as the Moon.

The size of the moon

Even though the Moon can be seen blazing brilliantly in the night sky and sometimes during the day, it may be difficult to comprehend exactly how big and how far away our closest neighbor truly is.

So how large is the Moon exactly?

It's not exactly as simple an answer as you may imagine. The Moon is not perfectly spherical, just like Earth. It's somewhat crushed instead. This indicates that the diameter of the Moon is less from pole to pole than it is at its equator.

However, the distance is just four kilometers apart. The Moon has a 3,476 km equatorial diameter and a 3,472 km polar diameter.

We must contrast it with something of comparable sizes, such as Australia, to understand how enormous that is.

between the coasts
Perth and Brisbane are 3,606 kilometers apart, as the crow flies. Australia and the Moon seem to be about the same size when placed next to one another.

But that's just one perspective on the situation. The Moon has almost the same width as Australia, but when you consider surface area, it is far larger. It turns out that Australia's surface is substantially smaller than the Moon's.

Australia's land area is about 7.69 million square kilometers. Contrarily, the Moon's surface measures 37.94 million square kilometers or almost five times the size of Australia.

How far away is the Moon?

Another issue whose solution is harder to figure out than you would think is how far away the Moon is.

As a result of the Moon's eccentric orbit around the Earth, its distance from our planet varies continually. The size of the Moon in our sky fluctuates a little from week to week since that distance may change by up to 50,000km in a single orbit.

Every other object in the Solar System has an impact on the Moon's orbit. Even after accounting for everything, the distance answer is still subject to change because of the Moon's progressive separation from the Earth due to tidal interaction.

The Apollo missions allowed us to more thoroughly investigate the final point. The Moon's surface was covered with a collection of mirror reflectors by the astronauts who traveled there. Lasers fired from the Earth are constantly aimed at those reflectors.

Scientists can measure the distance to the Moon with extreme precision and monitor the Moon's recession from Earth by timing how long it takes for that laser light to travel to and from the Moon. The outcome? 38mm per year, or just under 4 meters per century, is the rate at which the Moon is retraction.

Transport me to the moon

After all of that, the Moon and Earth are 384,402 kilometers apart on average. So let's explain what it means.

I would travel 4,310 kilometers (km) if I took the quickest Google-recommended route from Brisbane to Perth. Driving across the length of our nation would take around 46 hours.

I would need to go more than 89 times if I wanted to log enough kilometers to claim that I had traveled from Earth to the Moon. If I didn't encounter any traffic jams along the route, it would need five and a half months of continuous driving.

The Apollo 11 astronauts were fortunate not to be constrained by Australian speed constraints. After being launched on July 16, 1969, the command module Columbia took just three days and four hours to reach lunar orbit.

Eclipse-related coincidence

Nearly 1.4 million kilometers, or roughly 400 times the diameter of the Moon, make up the equatorial diameter of the Sun.

The fact that the distance between the Earth and the Sun (149.6 million kilometers) is almost (but not quite) 400 times that of the Moon results in one of astronomy's most amazing anomalies.

The outcome? In the sky above Earth, the Moon and the Sun seem to be about the same size. Because of this, a complete eclipse of the Sun occurs when the Moon and Sun coincide precisely, as viewed from Earth.

Sadly, Earth will never longer see such magnificent eclipses in the future. The Moon is receding and will soon be too far away to completely block the Sun. But that day won't come soon; according to most predictions, it won't happen for some 600 million years.

The space travelers

The Moon is still the only other globe where humans have set foot, despite sending robot envoys to the polar regions of the Solar System.

The number of persons who have stepped on the moon and are still living has dramatically decreased fifty years after the first expedition. That experience has been enjoyed by twelve individuals, but as of right now, just four are left.

These 12 moonwalkers just touched the surface of the Moon's enormous surface. Hopefully, we will come back in the next years to motivate a brand-new generation and carry on the human race's direct investigation of our closest planetary neighbor.

Chapter 3

Formation of the blue seas and oceans

Background of the Ocean

Around 4.5 billion years ago, when the Earth first began to form...
The ocean is not only an area of land where water is present. Geographically speaking, the ocean bottom differs from the continents. It governs a large portion of the geology and geological history of the continents and is caught in an endless cycle of creation and destruction that molds the ocean.

Geological activities that take place under the ocean's surface have an impact on both dry land and marine life. Ocean basin-forming processes take place gradually over tens and hundreds of millions of years. In this timeframe, when a human lifetime is but a blink of an eye, continents move over the surface of the globe, mountains emerge from level plains, and solid rocks flow like liquid. We must learn to embrace the alien perspective of geological time if we are to comprehend the ocean bottom. Marine biology heavily relies on geology. The areas where creatures reside, or their habitats, are directly formed by geological processes. This geology controls many aspects of a marine ecosystem, including the shape of the coasts, the depth of the water, and whether the bottom is muddy, sandy, or rocky. Paleontology is the term used to describe the geologic history of life.

Our planet is unusual because it has a lot of liquid water on it. On the majority of other worlds, there isn't much water, and when there is, it either exists as permanently ice or as vapor in the atmosphere. On the other hand, the earth is a water planet. The majority of the world's surface is covered by the ocean, which is essential in controlling our temperature and atmosphere. Life itself would not be possible without water.

72% of the earth's surface is made up of our ocean. Regarding the Equator, it is not spread uniformly. The Northern Hemisphere has around two-thirds of the planet's land area, which is just 61% ocean. Oceans cover over 80% of the Southern Hemisphere.

The ocean is often divided into four major basins. The biggest and deepest ocean is the Pacific, which is nearly as big as all the others put together. The Indian "Ocean" is somewhat bigger than the Atlantic "Ocean," although their average depths are comparable. The smallest and shallowest is the Arctic. Several shallow glasses of water, like the Mediterranean Sea, Gulf of Mexico, and the South China Sea, are connected to or marginal to the major ocean basins.

The seas are interrelated, even though we often think of them as four distinct entities. The globe map as viewed from the South Pole makes this the easiest to understand. The Pacific, Atlantic, and Indian seas are obvious big branches of one enormous ocean system from this vantage point. Seawater, minerals, and certain animals can flow from one "ocean" to another because of the links between the main basins. Oceanographers sometimes refer to one ocean as the "global ocean" since the "oceans" are essentially one large, linked system. They also call the body of water that encircles Antarctica by its continuous name, the Southern Ocean.

About 4.5 billion years ago, a dust cloud or clouds are assumed to have given birth to the planet and the rest of the solar system. Astrophysicists think that the big bang, which produced this dust, happened roughly 15 billion years ago, and left behind a lot of cosmic debris. As the dust fragments interacted, they combined to form bigger particles. As each of these bigger particles hit, pebble-sized rocks were created, which then collided to create larger boulders, and so on. As the process went on, the earth and other planets ultimately began to take shape.

The early earth created so much heat during its formation that it was likely molten. This made it possible for materials to settle on the globe following their densities. The mass or, more accurately, the weight of a given volume of a material is its density. Although an ounce of lead weighs less than a pound of styrofoam, most people consider lead to be "heavier" than styrofoam. This is because when lead and Styrofoam are compared using comparable amounts, lead weighs heavier. Lead is thus denser than Styrofoam. By dividing a material's mass by volume, the density of the substance may be determined. If two materials are combined, the less dense material will often float while the denser material will likely sink.

The densest material tended to flow toward the planet's core while lighter stuff floated toward the surface while the early earth was still in the molten state. A thin crust was formed as the thin surface material cooled. Oceans and the atmosphere eventually started to develop. The planet would have been so hot that all the water would have evaporated into the atmosphere if the earth had just slightly shifted its orbit closer to the sun. All the water would be permanently frozen in an orbit that is just slightly further from the sun. Fortunately for humans, a small region of our planet's orbit around the sun allows for the presence of liquid water. There wouldn't be any life on Earth if there wasn't liquid water.

The iron-rich core, the semi-plastic mantle, and the thin outer crust make up the earth's three primary layers. The earth's crust is the part we know the best. It is very thin when compared to the deeper layers, like hard skin floating on top of the mantle. Oceans and continents have quite different crusts in terms of composition and properties.

Oceans and continents vary geologically from one another due to the physical and chemical characteristics of the rocks, not because the rocks are submerged in water. Because of the makeup of the underlying rock, the ocean—the portion of the planet covered by water—is covered.
The sea bottom is made of oceanic crustal rocks, which are composed of dark-colored minerals known as basalt. The majority of continental rocks are of a form known as granite, which is typically lighter in color and has a distinct mineral composition than basalt. Although both are less thick than the underlying mantle, ocean crust is denser than continental crust. Similar to how icebergs float on water, the continents may be seen as large pieces of crust that are

"floating" on the mantle. The oceanic crust also floats on the mantle, but since it is denser than the continental crust, it does not float as high. This explains why the oceanic crust is below sea level and is submerged in water while the continents are high and dry above sea level. The geological ages of the continental and oceanic crusts are likewise different. Less than 200 million years old, or extremely young by geological standards, is the earliest oceanic crust. Conversely, continental rocks may be quite old—up to 3.8 billion years—!

After World War II, sonar made it possible to conduct the first thorough surveys of significant portions of the ocean bottom. The mid-oceanic ridge system, a 40,000-mile-long continuous chain of volcanic undersea mountains and valleys that round the globe like the seams of a baseball, was discovered as a consequence of these investigations. The greatest geological formation in the world is the mid-oceanic ridge system. The mid-ocean ridge is periodically shifted to one side or the other by transform faults, which are fissures in the crust of the planet. Rarely, do the ridge's undersea mountains reach such a height that they break the surface to create islands like Iceland and the Azores.

The Mid-Atlantic Ridge, which is a section of the mid-ocean ridge that spans across the Atlantic Ocean, closely follows the contours of the opposing coasts. The ridge rises to the eastern edge of the Pacific and creates an inverted Y in the Indian Ocean. The East Pacific Rise is the name of the major ridge in the eastern Pacific. Surveys also uncovered a network of trenches or deep depressions on the sea bottom. In the Pacific, trenches are extremely prevalent.

Geologists started closely examining the mid-ocean ridge system and trenches after its discovery because they were curious about how they were created. They discovered that these structures are surrounded by a lot of geological activity. For instance, volcanoes are more frequent in trenches while earthquakes are concentrated along ridges. The features of the rocks that make up the ocean bottom are connected to the mid-oceanic ridges. Rock samples from the real sea bottom were collected starting in 1968 by the Glomar Challenger, a deep-sea drilling ship. It was discovered that rocks became older the more away they are from the ridge crest. Investigating the magnetic properties of rocks on the ocean bottom led to one of the most significant discoveries. Parallel to the ridge is bands of rock with alternating normal and reversed magnetism.

In the end, an understanding of plate tectonics was brought about by the finding of the magnetic anomalies on the ocean bottom, together with other pieces of evidence. Many plates make up the earth's surface. The lithosphere is made up of these plates, which are made up of the crust and the uppermost layers of the mantle. There are roughly 100 kilometers of plates. The old lithosphere is destroyed someplace else while the new lithosphere is formed. In the alternative, the planet would need to keep growing to accommodate the new lithosphere. In the trenches, the lithosphere is destroyed. When two plates contact, one of the plates dips below the other and falls back down into the mantle, creating a trench. Subduction is the term for this downward motion of the plate into the mantle. Trenches are often referred to as subduction zones since it is where subduction takes place. Underwater volcanoes and earthquakes are both caused by subduction. Chains of volcanic islands might be formed by the volcanoes as they rise from the ocean bottom.

We now understand that there have been significant changes to the earth's surface. The ocean basins have altered in size and form, and the shifting sea bottom has transported the continents at great distances. In actuality, new oceans have formed. Scientists have been able to recreate a large portion of the history of these shifts because of their understanding of plate tectonics. For instance, researchers have shown that the continents were formerly part of a single supercontinent named Pangaea, which split apart 180 million years ago. Since then, the continents have relocated to their current location.

Seawater

The properties of saltwater are a result of the composition of the elements dissolved in it as well as the nature of pure water. There are two basic sources of solids that dissolve in saltwater. Some are created when rocks on land undergo chemical weathering, and rivers transport them to the ocean. Other materials originate deep inside the earth. At hydrothermal vents, the majority of these are expelled into the ocean. Some are ejected from volcanoes into the sky and reach the ocean as rain and snow. Although practically everything is present in seawater in some quantity, the majority of the solutes, or dissolved elements, are composed of a remarkably tiny group of ions. Almost 98% of the solids in saltwater are made up of only six ions. About 85% of the solids are made up of sodium and chloride, which is why saltwater has a flavor similar to table salt. The creatures that reside there are significantly impacted by the salinity of the water. For example, most marine species will perish in fresh water. Some species will be harmed by even small salinity changes.

Chapter 4

Beyond our universe.

One of the most compelling questions you could posit is this, which humanity has been pondering since, well, pretty much the beginning of time: What is beyond accepted bounds? What lies beyond the limits of our maps? What lies outside the boundary of the universe is the ultimate answer to this query.

The response is, um, complicated.

We must first define exactly what we mean by "universe" to respond to the question of what exists outside of it. There cannot be anything outside the universe if you take it to mean literally everything that could exist in all of space and time. Even if you imagine the universe to be a certain size and that there is something outside of that volume, the outside object must also be a part of the universe.

WHAT DOES THE UNIVERSE MEAN?

We must first explain precisely what we mean by "universe" to respond to the issue of what exists outside of it. There cannot be anything outside the universe if you consider it to imply absolutely everything that might exist in all of space and time. Even if you conceive the universe to have a certain size and that there is anything outside of that volume, the outside object must also be a part of the universe.

Even if the universe were to be a nameless, formless, and vacuum of nothingness, it would still be considered a thing and included in the list of "all the things," making it, by definition, a component of the universe.

You don't need to worry about this paradox if the cosmos has an endless size. Because the cosmos is all there is and has no edges, there is no exterior to even consider.

Yes, our visible portion of the cosmos has an exterior. Light moves at a finite speed and the universe has a finite age. Therefore, we haven't seen light from every galaxy in the universe's history. The observable universe is currently 90 billion light-years across. Beyond that line, there are probably a lot more haphazard stars and galaxies.

And after that? It's difficult to say.

Cosmologists are uncertain as to whether the cosmos is endlessly huge or just very massive. Astronomers instead examine the universe's curvature to determine its size.

We can learn about the universe's general form from its geometric curvature at enormous scales. The cosmos can be endless if it is completely flat mathematically. It has a limited volume if it is curved, like the surface of the Earth.

The cosmos seems to be nearly completely flat according to the most recent observations and measures of its curvature. This may lead you to believe that there are limitless universes. But it's not quite that easy. The universe need not be indefinitely large, even in the case of a flat one. Consider the surface of a cylinder as an illustration. It has a limited dimension while being mathematically flat because parallel lines placed on the surface stay parallel (this is one definition of "flatness"). It's possible that the cosmos is similarly flat and closed in on itself.

The existence of an edge or outside doesn't always follow from the fact that the cosmos is limited. Our three-dimensional cosmos could be a part of a greater, multidimensional structure. That is OK and does exist in certain unusual physics theories. However, we presently lack the means to test that, and it has little to no impact on the universe's regular functioning.

And even though it may give you a headache, the cosmos need not be embedded even if it has a limited volume.

You could see the cosmos as a huge ball packed with stars, galaxies, and several other fascinating astronomical things. You may picture what it seems like from the outside like an astronaut would see Earth from a peaceful orbit above.

However, that outside viewpoint is not necessary for the universe to exist. The cosmos is what it is. The definition of a three-dimensional world without necessitating an outside of that universe is completely mathematically self-consistent. You are pulling a mental prank on yourself when you see the cosmos as a ball floating in the void, which mathematics does not call for.

Of course, it seems implausible that there could be a limited world with no space outside of it. Not even "nothing" in the sense of an undefined, utterly undefined object in mathematics. In actuality, asking "What's beyond the cosmos" is like asking "What sound does purple make." Because you're attempting to mix two unrelated ideas, your query is absurd.

There may be an "outside" to our cosmos. Again, however, that need not be the case. There is nothing in mathematics that necessitates an outside to explain the cosmos.

Don't worry if everything seems difficult and confused. The goal of creating complex mathematics is to provide us with the tools necessary to understand ideas that are beyond our comprehension. One of the benefits of contemporary cosmology is that it enables us to investigate the unthinkable.

How vast is the cosmos?

How vast is the environment we live in? We have an answer from what we can see, but it's probably far greater than that.

How vast is the cosmos? One of astronomy's core issues is this one. We may determine a diameter by searching for the object's furthest visible point from Earth (and, thus, its oldest, given the speed of light).

Astronomers can peer back in time to the seconds immediately after the Big Bang thanks to developing technologies. This may give the impression that we can see the whole cosmos. But a lot of factors, such as the universe's form and expansion, affect how big the cosmos is.

Because of this, scientists are unable to determine the exact size of the universe, however, we may make guesses.

The most precise and comprehensive map(opens in new tab) of the universe's earliest light was published in 2013 by the European Satellite Agency's Planck space mission. The cosmos is 13.8 billion years old, according to the map. Planck used the cosmic microwave background to determine the age.

According to Charles Lawrence, the U.S. project scientist for the mission at NASA's Jet Propulsion Laboratory in Pasadena, California, "The cosmic microwave background light is a traveler from far away and long ago." "It teaches us about the whole history of our cosmos when it comes,"

This allows researchers to study an area of space located 13.8 billion light-years distant because of the relationship between distance and the speed of light. Astronomers on Earth may point their telescopes at 13.8 billion light-years in every direction, like a ship in an empty ocean, placing the planet within an observable sphere with a radius of 13.8 billion light-years. The crucial phrase here is "observable"; the sphere restricts what scientists can see, but not what is there.

The sphere is far bigger than it seems to be, despite having a diameter of over 28 billion light-years. The cosmos is expanding, as understood by scientists. As a result, even though astronomers may be able to make out a region that was 13.8 billion light-years away from Earth at the time of the Big Bang, the universe has continued to grow through time. According to Ethan Siegel, writing for Forbes(opens in new tab), assuming inflation happened at a constant pace throughout the history of the universe, the identical place is now 46 billion light-years distant, making the circumference of the observable universe a sphere of around 92 billion light-years.

The notion that the cosmos is not expanding uniformly makes these calculations more difficult. Data from NASA's Chandra Space Telescope, ESA's XMM-Newton, and Rosat X-ray observatories were used in a 2020 research that was announced by ESA(opens in new tab). This study reveals that the universe is not expanding uniformly in all directions. In this work, the brightness of hundreds of galaxy clusters was matched to their X-ray temperatures. Some clusters looked to be moving more slowly than anticipated since they weren't as brilliant as planned. According to ESA, "this potentially inconsistent impact on cosmic expansion may be brought about by the enigmatic dark energy."

Humans may seem to be at the center of the cosmos when a sphere is centered on the Earth's position in space. We are unable to determine our exact location inside the vast expanse of the cosmos, much as the ship in the ocean. Just as we cannot see the edge of the ocean or the limit of the cosmos does not necessarily indicate we are there, we cannot see land or the edge of the ocean does not necessarily mean we are there.

Numerous techniques have been used by scientists to gauge the size of the cosmos. They can detect the baryonic acoustic oscillations, which make up the cosmic microwave background and are waves from the early cosmos. They may also estimate distances using common candles, such as type 1A supernovae. However, the solutions may be found in these several distance measurement techniques.

It's unclear how inflation is evolving. Although the estimate of 92 billion light-years is based on the assumption that inflation is occurring at a constant pace, many scientists believe that the rate is decreasing. It should be 1023, or 100 sextillions if the universe expanded at the speed of light during inflation. The expansion of the universe in the seconds after the Big Bang may have been altered by dark energy events, according to one theory offered by NASA(opens in new tab) in 2019.

An Oxford University research team led by Mihran Vardanyan analyzed all of the data statistically rather than using just one measuring technique. Using Bayesian model averaging, which puts less emphasis on how well the model itself fits the data and more on how probable it is that it is accurate given the data. They discovered that the universe is at least 7 trillion light-years big, or at least 250 times larger than the visible cosmos.

According to a 2011 MIT Technology Review(opens in new tab) study, "That's large, but more tightly restricted than many other models."

the universe's shape?

The form of the cosmos has a significant impact on its size. The cosmos might be closed like a sphere, limitless and negatively curved like a saddle, or flat and infinite, according to scientific theories.

The form of the cosmos has a significant impact on its size. The cosmos might be closed like a sphere, limitless and negatively curved like a saddle, or flat and infinite, according to scientific theories.

Chapter 5

9 truths about extraterrestrials

The "alien debris" in our solar system, a shocking UFO story, and other fresh hints about extraterrestrial life.

In the sky above Switzerland, a "doughnut UFO." Over Canada, strange green lights disappear into the sky. A blob in the form of a saucer suddenly dives into the water.

Truth-seekers and extraterrestrial researchers have no lack of questions to consider in 2021. But it also provided them with answers, from a highly anticipated Pentagon study on military UFO encounters to fresh information on habitable exoplanets to the actual nature of a purported "alien signal" from the star that is closest to the sun. Here are nine facts we discovered about aliens in 2021 (and where to search).

1. UFOs exist

A highly anticipated study from the Pentagon covering 144 UFO sightings between 2004 and 2021 was published in June. Several UFO encounters that had, up until that point, only been reported via viral media were formally validated in the study, which was intended to analyze "the danger presented by unexplained aerial phenomenon (UAP)". On the one hand, the succinct 9-page study affirmed that "most of the UAP identified represent tangible things," ranging from birds and balloons to foreign spying technology and top-secret U.S. government operations. However, the report's failure to attribute any of the 144 encounters to extraterrestrial activity may have disappointed those looking for recognition of extraterrestrial intelligence.

2. Black holes may function as extraterrestrial power plants

While extraterrestrial explorers devote a lot of effort to looking for habitable planets outside of our solar system, research that appeared in the journal Monthly Notices of the Royal Astronomical Society in July cautions that black holes are among nature's most severe formations. Black holes may be attractive targets for extraterrestrial civilizations trying to fuel their interstellar endeavors since they may emit up to 100,000 times more energy than a star like our sun. Aliens may do this by radiating energy out into space using high-tech devices called Dyson spheres, which are enormous orbs that drain energy from the disc of white-hot matter spinning around a black hole's horizon. Dyson spheres were initially postulated in the 1960s. The authors of the researchers hypothesized that this re-radiated energy would produce a distinctive wavelength signature that astronomers might pick up from Earth. To scan through the data already collected by telescopes for such recognizable signs, the researchers are presently creating algorithms.

3. Aliens' worlds could not resemble Earth at all.

Research published in the Astrophysical Journal in August suggests that there may be another class of alien worlds that are just as hospitable to life as Earth-like planets, which is often where the hunt for extraterrestrial life starts. "Ocean" planets, which can be up to 2.5 times as massive as Earth and have vast oceans of liquid water below hydrogen-rich atmospheres, maybe the perfect home for microbes like "extremophiles," which can survive in some of Earth's most extreme environments (like hydrothermal vents), according to the study's authors. These planets are not only numerous in the Milky Way galaxy, but also highly varied, with some circling close to their home stars and others orbiting at great distances. According to the authors, both may be home to minute life below the surface, opening up a whole new field of investigation for those looking for extraterrestrial planets.

4. There is yet a chance for life on one of Saturn's moons.

Research published in June suggested that the methane emanating from Enceladus, Saturn's sixth biggest moon, might be an indication of the presence of life in the moon's deep water. Geysers spewing water ice into space were found in 2005 by NASA's Cassini Saturn probe, which was looking for "tiger stripe" cracks near the south pole of Enceladus. The orbiter discovered a wide range of other substances, including dihydrogen (H2) and a variety of carbon-containing organic compounds, including methane (CH4), as well as water. This material is considered to originate from a vast ocean of liquid water that sloshes under the moon's frozen surface.

Researchers in the latest study ran several simulations to see whether such molecules would be signs of microorganisms that "consume" dihydrogen and waste methane as methane. The scientists discovered that methane-farting bacteria may be causing the planet's gaseous geysers, suggesting that there may be life on the frozen moon.

5. "Alien garbage" in our solar system could be ignored by scientists.

The strange, cigar-shaped object 'Oumuamua, which zipped through our solar system in 2017, is almost certainly an example of alien technology, according to Harvard astrophysicist Avi Loeb's new book "Extraterrestrial: The First Sign of

Intelligent Life Beyond Earth" (published in January by Mariner Books). In his book, Loeb makes the case that 'Oumuamua is not naturally occurring, but rather an artifact of alien technology that may have been accidentally jettisoned into our solar system due to its unusual, elongated shape (unlike any known comet), extreme brightness, and apparent acceleration away from the sun.

"A buoy. A grid of communication pods The obsolete or abandoned technology of other intelligent living things, "Loeb penned. All of these answers for the "Oumuamua enigma" are reasonable ones since people currently practice them here on Earth, although on a much smaller scale. (Most scientists who have investigated the item prefer natural explanations; they refer to it as a cosmic "dust bunny" or simply a particularly strange comet.)

6. Thousands of extraterrestrial planets may have seen the development of mankind.

More than 1,700 extraterrestrial civilizations may have been keeping an eye on humanity for thousands of years earlier, even though human attempts to identify alien civilizations among the stars just began in the past century or so. Over the last 5,000 years, 1,715 neighboring star systems have had a perfect viewing angle of Earth, and more than 1,400 of them still do, according to research published in June in the journal Nature(opens in new tab).

All of these stars are located within 300 light-years of Earth, and 75 of them have orbital distances of less than 100 light-years. Any of those 75 star systems are close enough that "our radio waves would have washed over them already," according to lead study author Lisa Kaltenegger, an associate professor of astronomy and director of the Carl Sagan Institute at Cornell University, who told Live Science at the time. Humans have been sending radio signals for about 100 years. Another concern is whether any potential civilizations inhabiting those star systems would want to interact with us.

7. There is no "optimal" method of contacting aliens.

What would be the greatest approach to let aliens know where we live if they were monitoring us from a close distance? Joanna Thompson, a reporter for Live Science, looked into this issue in December and discovered that no approach is perfect. On the one hand, radio waves are an alluring means of communication with extraterrestrials because they pass through the electromagnetic spectrum's "water hole," or the frequency range between 1420 and 1720 megahertz, which is largely free of cosmic background noise.

But as radio waves spread out as they go, whatever message we transmit will become diluted the farther away from Earth it is. Although laser communications need to be very precise and are unlikely to reach any extraterrestrial observers unless we direct our message precisely to their star system, laser light does not have this issue. Although neither approach is ideal, each has its merits.

8. Our technology may be impeding progress.

Astronomers discovered a signal on April 29, 2019, that seemed to be coming from Proxima Centauri, the closest star system to our sun and the location of at least one possibly habitable planet. Researchers hypothesized that the signal may have been produced by extraterrestrial technology since it fell into a certain radio wave spectrum that is seldom produced by human satellites or airplanes. But the signal never reproduced — and research published in October in the journal Nature Astronomy(opens in new tab) explains why: The telescope that picked up the signal was actually near a malfunctioning computer or mobile device that was generating the signal.

The 2019 data were reexamined in the new study, and the researchers discovered several "lookalike" signals that appeared to be missing pieces of the alleged alien transmission. Taken together, these signals fit a range of frequencies "consistent with common clock oscillator frequencies used in digital electronics," the researchers wrote. In other words, this extraterrestrial communication seems to have been a human computer malfunction, but analyzing and recognizing it still provides researchers useful practice in distinguishing between genuine deep-space signals and Earthly noise.

9. "Abductions" by aliens could be lucid dreams.

The researchers asked 152 lucid dreamers to have dreams about coming across aliens or flying objects and discovered that many of the dreamers had visions that mirrored reports of purported alien abductions. 24 percent of individuals who said their dream interactions were "realistic" also reported having severe dread and sleep paralysis. Although people who claim to have been abducted by aliens may sincerely believe that what they experienced was real, the study's authors noted that these people were likely having an extraterrestrial encounter while in a lucid dream. Such emotions frequently go along with reports of alleged alien abductions.

www.ingramcontent.com/pod-product-compliance
Lightning Source LLC
Chambersburg PA
CBHW070302220526
45465CB00004B/1710